CAMBRIDGE
POCKET
STAR FINDER

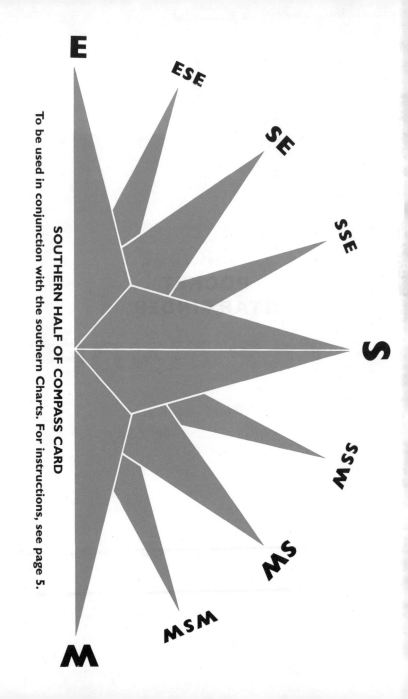

SOUTHERN HALF OF COMPASS CARD

To be used in conjunction with the southern Charts. For instructions, see page 5.

CAMBRIDGE
POCKET
STAR FINDER

A month-by-month
guide to the night sky

CAMBRIDGE
UNIVERSITY PRESS

Published by the Press Syndicate of the University of Cambridge
The Pitt Building, Trumpington Street, Cambridge CB2 1RP
40 West 20th Street, New York, NY 10011-4211, USA
10 Stamford Road, Oakleigh, Melbourne 3166, Australia

First published in 1918 as *Stars at a Glance*
Completely revised © 1959 and 1970
Retitled *Star Finder* 1991
Revised 1995

Published in Great Britain in 1995 by George Philip Limited

First published by Cambridge University Press 1996

Library of Congress cataloguing in publication data available

This edition only for sale in the United States of America and Canada

ISBN 0 521 58993 2 paperback

Printed in Hong Kong

CONTENTS

Introduction to the study of the stars

Two facts are at once evident to a careful observer of the night sky: (1) that the position of the stars is fixed relative to one another, and (2) that they appear to rotate round a central point near the star Polaris. This star is nearly overhead at the North Pole and, in relation to it, all other stars appear to follow parallel paths in a series of concentric circles called parallels of declination.

Apparent rotation of the stars

From North America Polaris always appears due north, midway between the zenith, which is directly overhead, and the northern horizon. From the northern USA, therefore, the parallel paths followed by the stars round Polaris are inclined at an angle to our horizon, the angle always being equal to the latitude of the observer. It follows that stars which are always low on the horizon at the North Pole appear to us to rise on our eastern horizon. After rising, they ascend in the sky until they cross the southern meridian, and then descend until they set on the western horizon, and finally disappear for the remainder of the journey along their circular pathway until they re-emerge the following night. Because of this tilt, many more stars are visible in the USA than at the North Pole, but, again for the same reason, not all at the same time.

In their rotation about the Pole Star, certain stars never pass below the horizon and are known as northern circumpolar stars (see page 9). On the contrary, the southern circumpolar stars, being always below our horizon, are never visible to us (see page 40). Both by night and day, the stars are pursuing their steady course overhead, though in daylight they are unseen because of the brilliance of the Sun.

A complete rotation of the stars occupies 23 hours 56 minutes, which is the length of sidereal day, as opposed to the solar day of 24 hours. A star will therefore be seen in the same position in the night sky four minutes earlier each succeeding night, or about two hours earlier every month.

The effect of this is illustrated quite clearly by following the change in position of any constellation on the monthly Charts (see pages 12–35). For example, follow the path of Orion on the southern Charts, from October to March. This gain of four minutes a day on solar time results in an advance of about 24 hours in the year. Thus, any constellation returns to its original starting point in a year of 365 days, after making a complete revolution. It should be noted that in addition to showing the position of the stars for every month of the year, the charts also illustrate their position for every two hours of the night.

Magnitude

Stars are classified into magnitudes according to their apparent brightness; that is, the amount of light apparently emitted. This qualification is important because a faint star may appear to be emitting less light than another merely because it is so much more distant. First-magnitude stars are the most brilliant, and are two and a half times brighter than second-magnitude stars, which in turn are two and a half times brighter than third-magnitude stars, and so on. Magnitudes are shown by symbols on the charts. Recognition by magnitude is not very satisfactory, since the seasonable position of any one star may be the reason for considerable dimming due to the additional atmosphere through which it is viewed. Moreover, there may be local haze due to weather. The so-called variable stars appear to change their brilliancy and are represented on the charts by their average magnitude. They are generally of more interest to the regular observer than to the casual stargazer.

The Milky Way

The Milky Way (omitted from the charts for the sake of greater clarity) is an indistinct luminous band stretching right across the heavens, and is formed by the concentrated light of countless stars that lie between us and the outer rim of the Galaxy which is our home. Under first-class conditions it is a striking phenomenon and is a useful aid in recognizing groups which lie on or near to it.

Constellations

Ancient astronomers arbitrarily separated the stars into groups, now known as constellations, and each group was conceived to represent some mythological figure and was named accordingly. The principal constellations are described on pages 41–48. The stars forming a constellation are rarely together in fact, but only appear so in our line of sight. Those of the Big Dipper are an exception to this rule. Clusters, however, are compact groups travelling together, and are indicated on the charts by special symbols. Nebulae are of several kinds, under the two main groupings of gaseous nebulae – cloudy masses of luminous gas – and extra-galactic nebulae, which are thought to be complete island universes so distant that the separate stars composing them are indistinguishable.

Planets

Planets are satellites of the Sun, depending on the Sun for heat and shining by reason of its reflected light. They revolve in regular and clearly defined orbits around the Sun, but, relative to the fixed stars, their motion is irregular and so it is not possible to show them on the charts.

Their steady radiance is readily distinguished from the "twinkle" of the stars.

In addition to the notes on page 11, the reader is referred to the listings published in some newspapers, astronomical magazines or yearbooks for the dates and times of the appearance of the planets. Mercury and Venus are nearer the Sun than the Earth is, and therefore only appear as morning or evening stars (see page 50 for planetary distances from the Sun). Venus is the next brightest object in our heavens after the Sun and Moon.

The Sun, Moon and planets all follow paths across the sky within a definite zone which has been named the zodiac, and the zodiacal constellations are those lying within this zone. The path of the Sun is called the ecliptic and is in the middle of this zone. Owing to the tilt of the Earth's axis, the ecliptic crosses the equator at two points, known as the spring and fall equinoxes (see pages 17 and 29).

Mapping the stars

Although the distances of the stars from the Earth vary enormously, to us they appear to be on the inside of a huge sphere enclosing the Earth. Mapping them offers the same problem of projection as mapping the Earth: that of representing a sphere, in this case the apparent sphere of the heavens, on the flat surface of a map. In mapping the celestial sphere, North and South Poles are considered to be directly above the North and South Poles of the Earth. This same principle also governs the position of the celestial equator or equinoctial. Parallels of latitude, however, become parallels of declination, and meridians of longitude are called hour circles of right ascension, usually abbreviated to "Decl." and "RA." The sidereal hour angle (SHA) is represented by $360° - RA$, measured from the "First Point of Aries," i.e. the point where the ecliptic intersects the equinoctial on March 21 each year.

In preparing a flat copy of what is apparently a sphere, it is impossible to avoid distortion at the edges of the chart. However, it has been kept to a minimum in this book by giving charts of the north and south on a polar projection (pages 9 and 40) and the middle heavens (that part directly over the equator) on a Mercator projection (pages 36–39). Any distortion is thus kept to the edges of the charts. All the stars which are included in the planisphere chart on pages 10–11 can be seen at some time from the northern USA.

Charts

The Charts are drawn on a system, peculiar to this book, intended to simplify recognition. The lines indicate direction only, and do not in any way correspond with declination and right ascension. Although each double-page Chart forms a complete circle, it must be clearly understood that it comprises two semi-circular pictures of the heavens – one looking north and one looking south – and the bottom of each picture is that edge next to the binding. The zenith for the northern USA falls midway along the curved edge of each chart. From it lines are drawn to the points of the compass on the horizon. Thus, the whole of the sky visible at the time for which the chart is drawn is divided into imaginary sectors. When comparing the chart with the sky, corresponding lines can be imagined in the sky which will serve as guidelines in identifying stars.

Compass cards

Just inside the covers of the book is a boldly designed compass card, the southern portion at the front and the northern portion at the back, which is clear enough to be read in dim light. The quarter points on it correspond with the lines on the Charts. By holding the unwanted pages vertically between the thumb and forefinger so that the edge is toward the observer, it is easy to arrange for any Chart to be viewed at the same time as its appropriate half of the compass card. If at first the observer always uses the northern chart and compass card, and fixes his/her position by Polaris, he/she can afterward face the reverse direction and use the southern chart and compass card. After the first few observations, it will be possible to identify the constellations visible with little preliminary orientation.

Only the chief groups are given on the Charts, together with their popular names. Once these have become familiar, the less conspicuous groups may be sought, and instructions to aid their identification are included in the catalog of stars on pages 41–48, which can be used in conjunction with the more detailed plates on pages 9 and 36–39.

Finding and naming stars without a compass

The simplest and most reliable guide to the true north direction is the **Pole Star (Polaris)**. An observer looking toward this star faces north (N), has his/her back to the south (S), and by extending his/her arms in line with the shoulders, obtains approximately west (W) and east (E) bearings. The relative positions and appearance of the following circumpolar guide stars should be memorized by means of the chart on page 9.

1. **Dubhe** and **Merak**, "the Pointers" (Big Dipper), are in a straight line with the Pole. The **Pole Star** lies a little to one side of this line.

2. **Caph** (Cassiopeia) and **Megrez** (Big Dipper): the North Celestial Pole is at the center of a straight line joining these stars.

3. **Vega** (Lyra) and **Capella** (Auriga) lie at a considerable distance on either side of the Pole, almost at right angles. **Polaris** appears to lie about halfway along a slightly curved line between them.

The first is the best method of finding the North Star, since the Big Dipper cannot be mistaken, and it is always high enough in the sky to be easily seen above trees in uneven country.

When the north is found, the remaining points can be fairly accurately obtained by means of the compass card provided inside the covers of the book. Hold the northern half so that the point N is directly under the Pole Star. The card is then a true compass, and the bearings from N to E and W can be read off. Points at the horizon bearing due E and W should be noted, and the base of the southern half of the card aligned with them. It can then be used in the same way as the northern half, the observer now facing south. The two portions of the card are so arranged as to be easily used in conjunction with the northern and southern Charts respectively.

To identify a star or constellation visible at a given time:
1. With the aid of the Calendar Index (page 7), select the Chart appropriate to the time and date of observation.

2. Imagine a straight line extending from the zenith, through the star, and by means of the compass card and Pole Star take the bearing of the point where the line reaches the horizon.

3. Find on the Chart the nearest bearing and guideline (here curved) when the bright stars and groups near the imaginary line in the sky can be identified.

To find a certain star in the sky:
1. Find the star on the Chart for the time of observation, and note which of the guidelines it is nearest.

2. Take the bearing of this line at the horizon with the aid of the compass card and Pole Star, and imagine a straight line joining it to the zenith.

3. Read off the most conspicuous stars near this line from the chart, identifying the one required by its elevation or position in regard to other stars.

Calendar Index to the monthly Charts (pages 12–35)

Each pair of charts shows the principal stars visible from 9 pm to midnight during the month. To ascertain the appropriate charts for a particular time and date, find the time at the head of the diagram and follow the vertical column downward to a point opposite to the required date. The diagonal band in which this point lies will indicate the charts to be consulted.

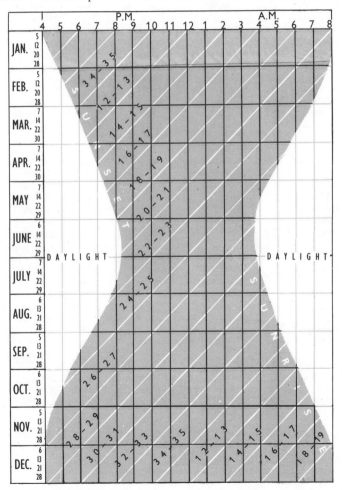

Air navigation stars

The following 57 stars appear in the Air Almanac. The list comprises three groups, the first one of 22 stars being the most important. Mention should also be made of Polaris (magnitude 2.1, Ursa Minor), a very important star for navigators.

Star Name	Mag.	Constellation	Star Name	Mag.	Constellation
Achernar	0.6	Eridanus	Dubhe	1.9	Ursa Major
Acrux	1.0	Crux	Fomalhaut	1.3	Piscis Australis
Aldebaran	1.1	Taurus	Peacock	2.1	Pavo
Alpheratz	2.1	Andromeda	Pollux	1.2	Gemini
Altair	0.9	Aquila	Procyon	0.5	Canis Minor
Antares	1.2	Scorpius	Regulus	1.3	Leo
Arcturus	0.2	Boötes	Rigel	0.3	Orion
Betelgeuse	0.1 to 1.2	Orion	Rigil Kentaurus	0.1	Centaurus
Canopus	−0.9	Carina	Sirius	−1.6	Canis Major
Capella	0.2	Auriga	Spica	1.2	Virgo
Deneb	1.3	Cygnus	Vega	0.1	Lyra

Star Name	Mag.	Constellation	Star Name	Mag.	Constellation
Alioth	1.7	Ursa Major	Kochab	2.2	Ursa Minor
Alkaid	1.9	Ursa Major	Nunki	2.1	Sagittarius
Alphard	2.2	Hydra	Rasalhague	2.1	Ophiuchus
Denebola	2.2	Leo	Schedar	2.5	Cassiopeia
Diphda	2.2	Cetus	Shaula	1.7	Scorpius
Hamal	2.2	Aries	Suhail	2.2	Carina

Star Name	Mag.	Constellation	Star Name	Mag.	Constellation
Acamar	3.1	Eridanus	Enif	2.5	Pegasus
Adhara	1.6	Canis Major	Gacrux	1.6	Crux
Al Na'ir	2.2	Grus	Gienah	2.8	Pegasus
Alnilam	1.8	Orion	Hadar	0.9	Centaurus
Alphecca	2.3	Corona Borealis	Kaus Australis	2.0	Sagittarius
Ankaa	2.4	Phoenix	Markab	2.6	Pegasus
Atria	1.9	Triangulum Australe	Menkar	2.8	Cetus
			Menkent	2.3	Centaurus
Avior	1.7	Carina	Miaplacidus	1.8	Carina
Bellatrix	1.7	Orion	Mirfak	1.9	Perseus
Elnath	1.8	Taurus	Sabik	2.6	Ophiuchus
Eltanin	2.4	Draco	Zuben'ubi	2.9	Libra

Reference to many of these stars is made in the section at the end of this book (pages 41–48), describing the constellations and giving hints on their location and other interesting facts.

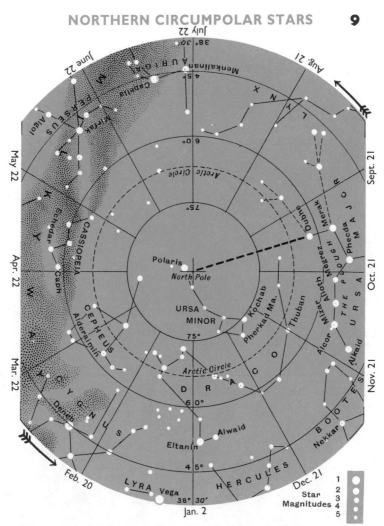

The stars shown above are (theoretically) *always visible* from or north
of the latitude 51.5°N. The direction of their apparent revolution is
indicated by arrows. The dates correspond with those of the Charts
(see pages 12–35) and show when the RA circles touch the north point
of the horizon at 10 pm, the distance between them representing two
hours' motion.

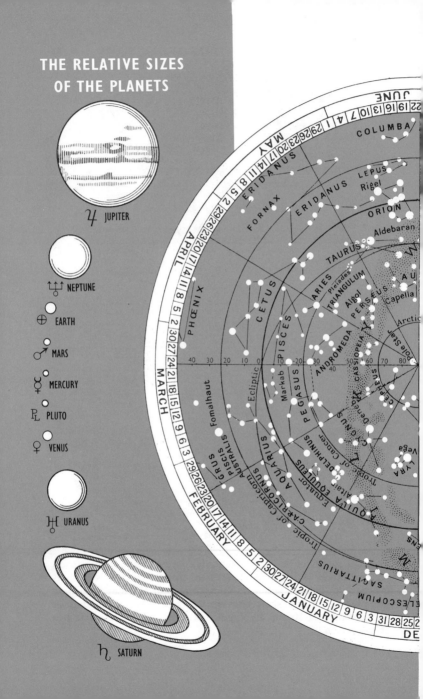

THE RELATIVE SIZES
OF THE PLANETS

♃ JUPITER

↑↑↑ NEPTUNE

⊕ EARTH

♂ MARS

☿ MERCURY

♇ PLUTO

♀ VENUS

♅ URANUS

♄ SATURN

STARS OF THE NORTHERN HEMISPHERE

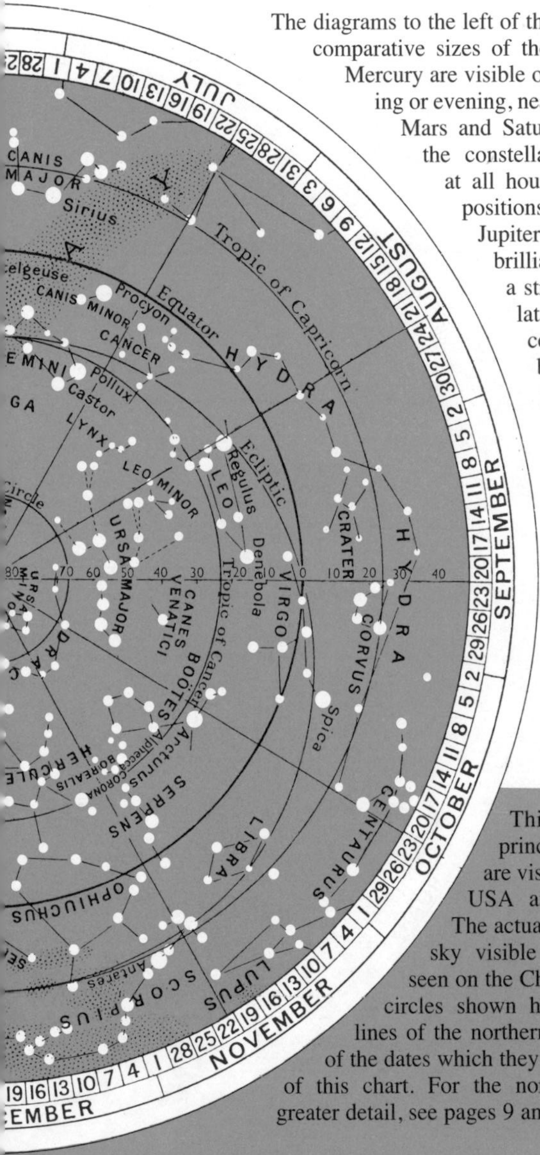

The diagrams to the left of this star chart show the comparative sizes of the planets. Venus and Mercury are visible only at twilight, morning or evening, near the ecliptic. Jupiter, Mars and Saturn are to be seen in the constellations of the zodiac at all hours of the night, their positions continually varying. Jupiter and Mars are equally brilliant, the former having a strong yellow light, the latter being reddish in colour. Saturn is less bright, but its steady yellow radiance helps to distinguish it from bright stars. Uranus, Neptune and Pluto are invisible to the unaided eye.

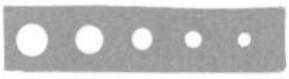

This chart shows the principal fixed stars that are visible from the northern USA and similar latitudes. The actual portions of the night sky visible at any date can be seen on the Charts. The 12 RA hour circles shown here form the middle lines of the northern Charts for midnight of the dates which they subtend at the margin of this chart. For the northern hemisphere in greater detail, see pages 9 and 36–39.

The principal January stars are shown here and on the opposite page. Their positions are correct for January 5 at 11 pm and January 20 at 10 pm, and for four minutes earlier on each succeeding night, for example January 1 at 11.16 pm, January 6 at 10.56 pm, January 21 at 9.56 pm, etc.

This view of the heavens is also correct for December 21 at midnight; February 5 at 9 pm; February 20 at 8 pm; March 7 at 7 pm.

The stars move from the positions shown here to those on the next pair of Charts in the space of two hours.

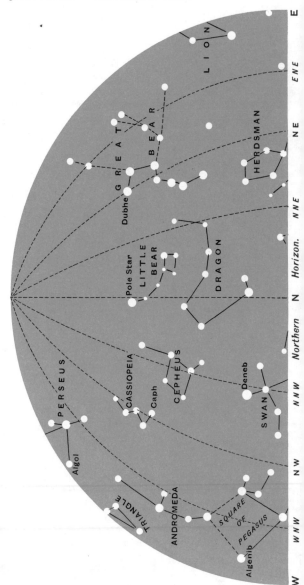

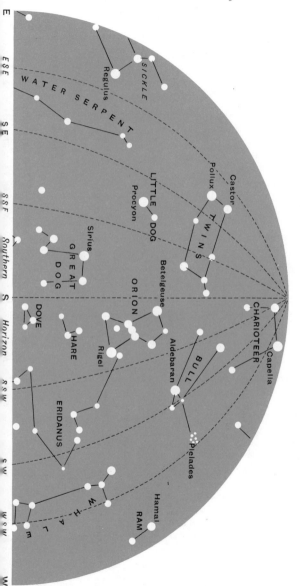

On January 1, day breaks at 6h 2m and sunrise is at 8h 8m am. The Sun sets at 4h and twilight ends at 6h 4m pm, the length of actual daylight being about 7 hours 50 minutes. During the month, the mornings increase by about 25 minutes and the afternoons by 45 minutes.

The principal February stars are shown here and on the opposite page. Their positions are correct for February 5 at 11 pm and February 20 at 10 pm, and for four minutes earlier on each succeeding night, for example February 1 at 11.16 pm, February 6 at 10.56 pm, February 21 at 9.56 pm, etc.

This view of the heavens is also correct for January 20 at midnight; March 7 at 9 pm; March 22 at 8 pm.

The stars move from the positions shown here to those on the next pair of Charts in the space of two hours.

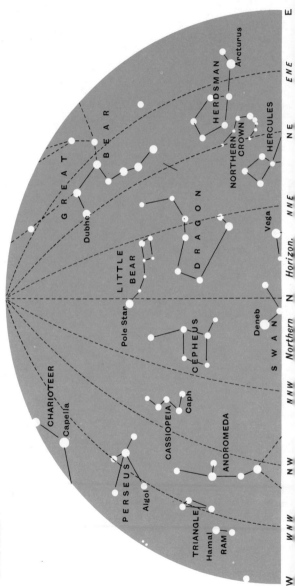

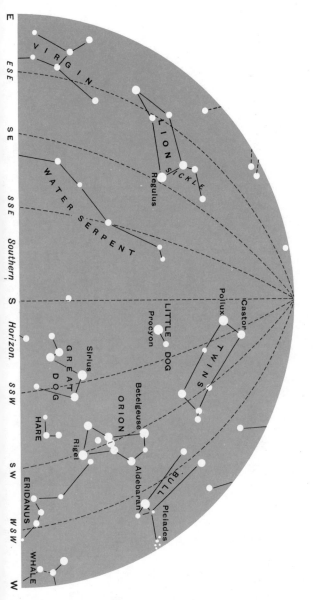

On February 1, day breaks at 5h 50m and sunrise is at 7h 42m am. The Sun sets at 4h 45m and twilight ends at 6h 45m pm, the length of actual daylight being about 9 hours.

The principal March stars are shown here and on the opposite page. Their positions are correct for March 7 at 11 pm and March 22 at 10 pm, and for four minutes earlier on each succeeding night, for example March 8 at 10.56 pm, March 9 at 10.52 pm, March 10 at 10.48 pm, etc.

This aspect of the heavens is also correct for February 20 at midnight; April 7 at 9 pm.

The stars move from the positions shown here to those on the next pair of Charts in the space of two hours.

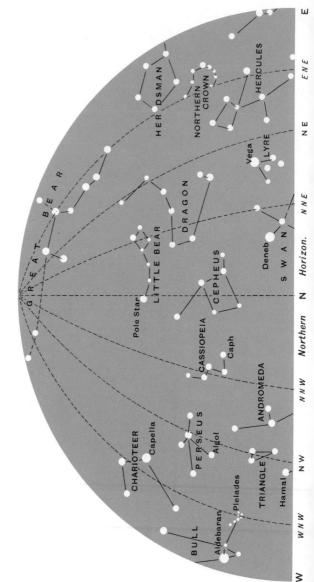

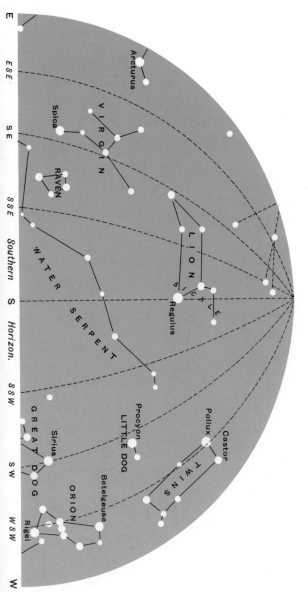

The point on the ecliptic between Pisces and Aries is known as the "First Point of Aries," and from it both the signs and right ascension hour circles are numbered. The spring equinox occurs when the Sun reaches this position, and day and night are equal all over the world, the Sun rising and setting due east and west at approximately 6 am and 6 pm.

The principal April stars are shown here and on the opposite page. Their positions are correct for April 7 at 11 pm and April 22 at 10 pm, and for four minutes earlier on each succeeding night, for example April 1 at 11.28 pm, April 2 at 11.24 pm, April 8 at 10.56 pm, April 23 at 9.56 pm, etc.

This view of the heavens is also correct for March 22 at midnight.

The stars move from the positions shown here to those on the next pair of Charts in the space of two hours.

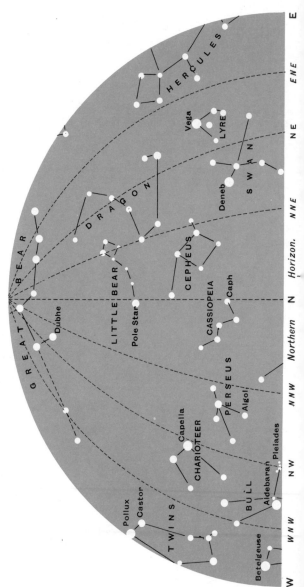

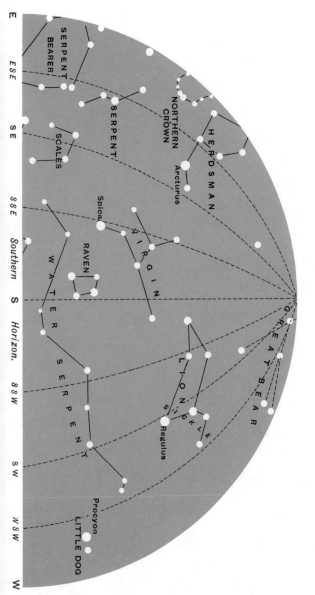

Days are now longer than nights. On April 1, day breaks at 3h 35m and sunrise is at 5h 38m am. The Sun sets at 6h 30m and twilight ends at 8h 30m pm, the period of actual daylight being about 13 hours. During April, the mornings increase by 62 minutes and the afternoons by 47 minutes.

The principal May stars are shown here and on the opposite page. Their positions are correct for May 7 at 11 pm and May 22 at 10 pm, and for four minutes earlier on each succeeding night, for example May 8 at 10.56 pm, May 9 at 10.52 pm, May 10 at 10.48 pm, May 23 at 9.56 pm, etc.

This view of the heavens is also correct for April 22 at midnight.

The stars move from the positions shown here to those on the next pair of Charts in the space of two hours.

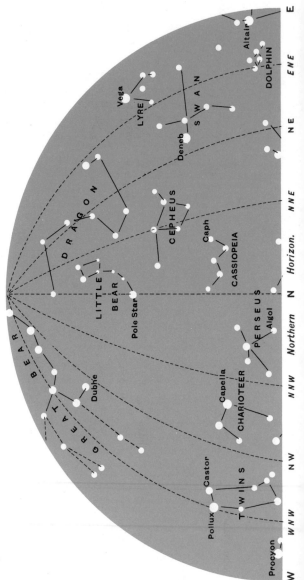

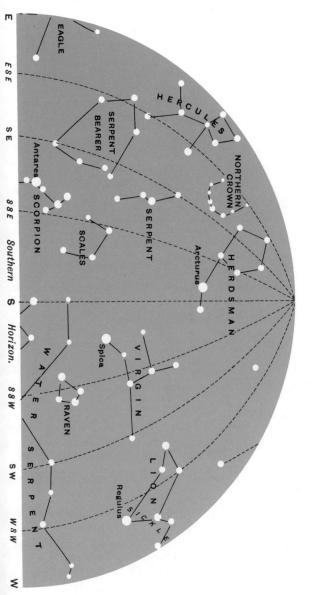

On May 1, daybreak is at 2h and the Sun rises at 4h 35m am. Sunset is at 7h 20m and twilight ends at 9h 50m pm, the period of actual daylight being about 14 hours 45 minutes. During May, the mornings increase by 44 minutes and the afternoons by 44 minutes.

JUNE – Northern View

The principal June stars are shown here and on the opposite page. Their positions are correct for June 6 at 11 pm and June 22 at 10 pm, and for four minutes earlier on each succeeding night, for example June 7 at 10.56 pm, June 8 at 10.52 pm, June 9 at 10.48 pm, June 23 at 9.56 pm, etc.

This view of the heavens is also correct for May 22 at midnight.

The stars move from the positions shown here to those on the next pair of Charts in the space of two hours.

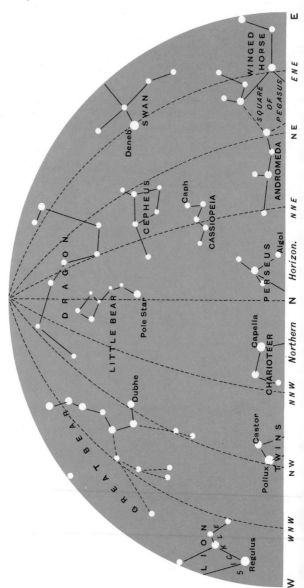

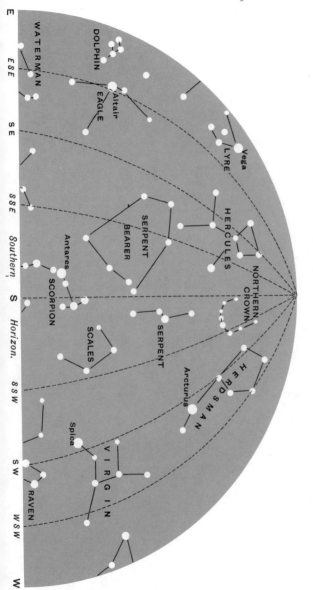

The Sun reaches the summer solstice about June 22, after which date the length of the day diminishes. At the solstice, the Sun rises and sets in the northeast and northwest. The short nights of June are wholly twilight, there being no real darkness. On June 1, the Sun rises at 3h 50m am and sets at 8h 5m pm, the length of day being about 16 hours 15 minutes.

The principal July stars are shown here and on the opposite page. Their positions are correct for July 7 at 11 pm and July 22 at 10 pm, and for four minutes earlier on each succeeding night, for example July 8 at 10.56 pm, July 23 at 9.56 pm, July 24 at 9.52 pm, July 25 at 9.48 pm, etc.

This view of the heavens is also correct for June 22 at midnight.

The stars move from the positions shown here to those on the next pair of Charts in the space of two hours.

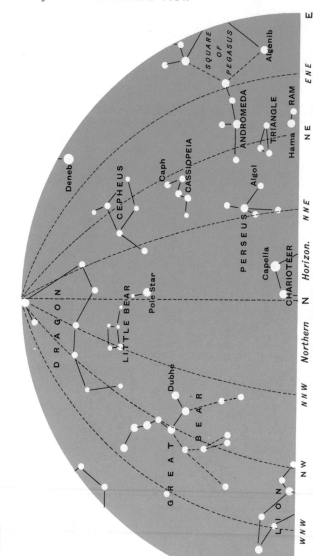

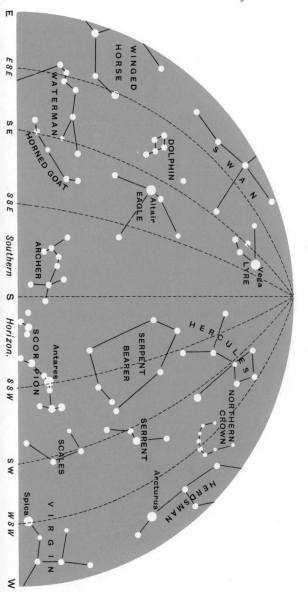

Until about the 21st of the month, there is no real darkness. On July 1, sunrise is at 3h 50m am and sunset at 8h 20m pm, the length of day being about 16 hours 30 minutes. The period of actual daylight now diminishes, the mornings decreasing by 34 minutes and the afternoons by 30 minutes during the month.

The principal August stars are shown here and on the opposite page. Their positions are correct for August 6 at 11 pm and August 21 at 10 pm, and for four minutes earlier on each succeeding night, for example August 22 at 9.56 pm, August 23 at 9.52 pm, August 24 at 9.48 pm, August 25 at 9.44 pm, etc.

This view of the heavens is also correct for July 22 at midnight; September 5 at 9 pm; September 21 at 8 pm; October 6 at 7 pm; October 21 at 6 pm.

The stars move from the positions shown here to those on the next pair of Charts in the space of two hours.

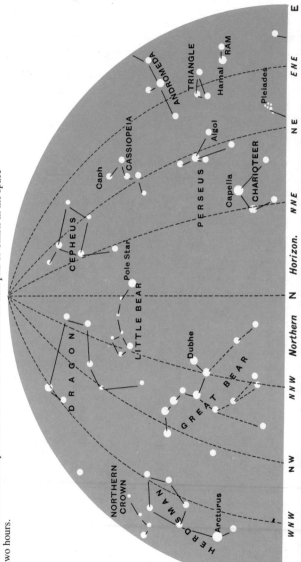

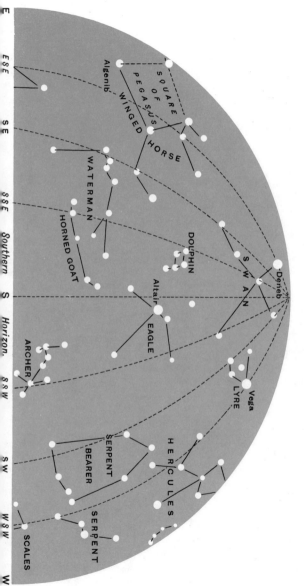

On August 1, day breaks at 1h 30m and sunrise is at 4h 25m am. The Sun sets about 7h 45m and twilight ends at 10h 40m pm, the length of actual daylight being about 15 hours 20 minutes. During the month, the mornings decrease by 47 minutes and the afternoons by 59 minutes.

The principal September stars are shown here and on the opposite page. Their positions are correct for September 5 at 11 pm and September 21 at 10 pm, and for four minutes earlier on each succeeding night, for example September 1 at 11.16 pm, September 6 at 10.56 pm, September 22 at 9.56 pm, etc.

This view of the heavens is also correct for August 21 at midnight; October 6 at 9 pm; October 21 at 8 pm, November 5 at 7 pm; November 21 at 6 pm.

The stars move from the positions shown here to those on the next pair of Charts in the space of two hours.

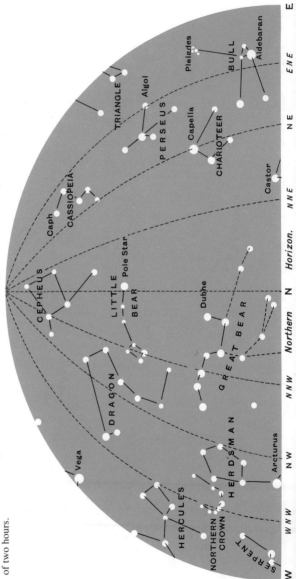

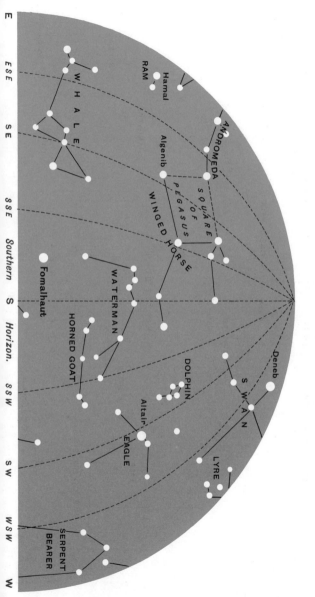

The autumnal equinox occurs about September 22, and day and night are equal all over the world. Fall now begins. On September 1, daybreak is at 3h 5m and the Sun rises at 5h 15m am. Sunset is at 6h 45m and twilight ends at 8h 50m pm, the period of actual daylight being about 13 hours 30 minutes. On September 22, the Sun rises and sets due east and west. During the month, the mornings decrease by 46 minutes and the afternoons by 66 minutes.

The principal October stars are shown here and on the opposite page. Their positions are correct for October 6 at 11 pm and October 21 at 10 pm, and for four minutes earlier on each succeeding night, for example October 1 at 11.20 pm, October 7 at 10.56 pm, October 22 at 9.56 pm, etc.

This view of the heavens is also correct for September 21 at midnight; November 5 at 9 pm; November 21 at 8 pm; December 6 at 7 pm; December 21 at 6 pm.

The stars move from the positions shown here to those on the next pair of Charts in the space of two hours.

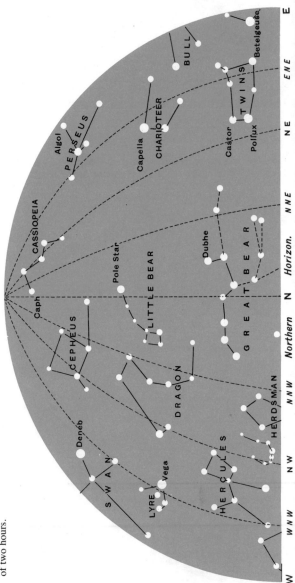

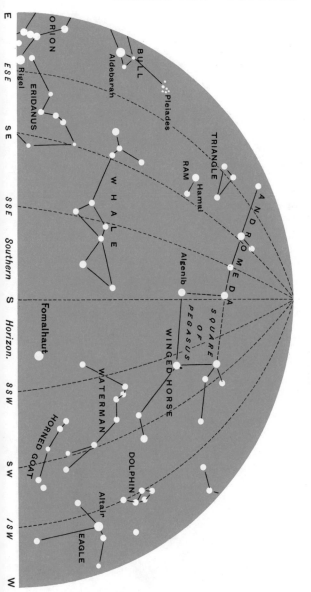

On October 1, daybreak is at 4h 10m and the Sun rises at 6 am. Sunset is at 5h 38m and twilight ends at 7h 30m pm, the length of actual daylight being about 11 hours 40 minutes. During the month, the mornings decrease by 51 minutes and the afternoons by 64 minutes.

The principal November stars are shown here and on the opposite page. Their positions are correct for November 5 at 11 pm and November 21 at 10 pm, and for four minutes earlier on each succeeding night, for example November 1 at 11.16 pm, November 6 at 10.56 pm, November 22 at 9.56 pm, etc.

This view of the heavens is also correct for October 21 at midnight; December 6 at 9 pm; December 21 at 8 pm; January 5 at 7 pm; January 20 at 6 pm.

The stars move from the positions shown here to those on the next pair of Charts in the space of two hours.

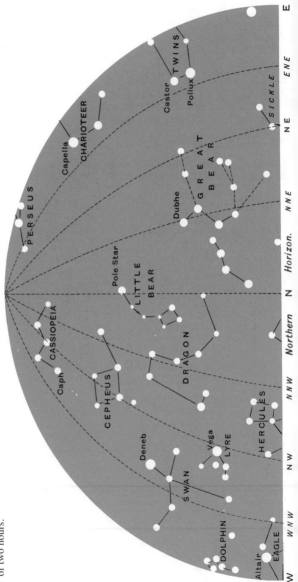

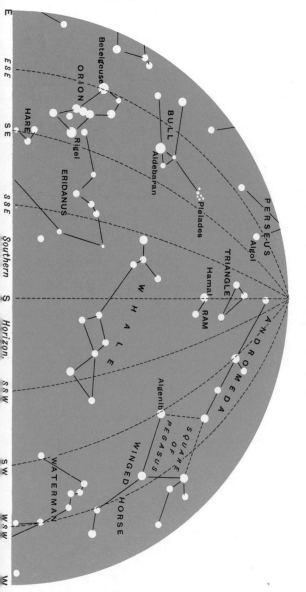

On November 1, day breaks at 5h and sunrise is at 6h 55m am. The Sun sets at 4h 35m and twilight ends at 6h 25m pm, the length of actual daylight being about 9 hours 40 minutes. During the month, the mornings decrease by 49 minutes and the afternoons by 40 minutes.

The principal December stars are shown here and on the opposite page. Their positions are correct for December 6 at 11 pm and December 21 at 10 pm, and for four minutes earlier on each succeeding night, for example December 1 at 11.20 pm, December 7 at 10.56 pm, December 22 at 9.56 pm, etc.

This aspect of the heavens is also correct for November 21 at midnight; January 5 at 9 pm; January 20 at 8 pm; February 5 at 7 pm.

The stars move from the positions shown here to those on the January Chart (see pages 12–13) in the space of two hours.

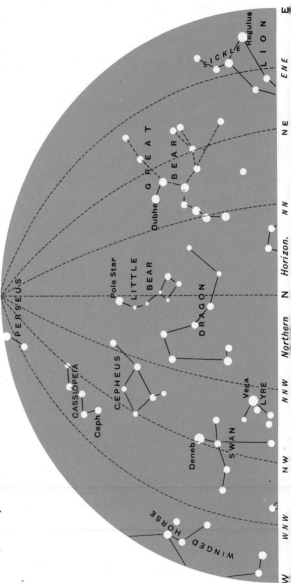

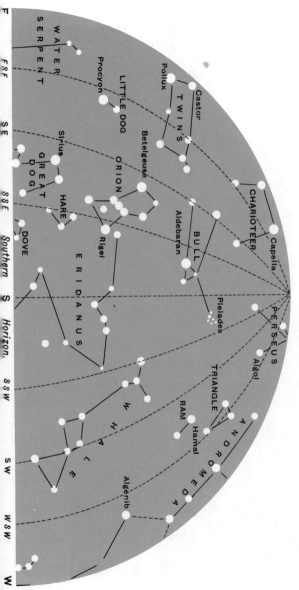

The Sun reaches the winter solstice on December 22, after which the days begin to lengthen. At the solstice, the Sun rises and sets in the southeast and southwest. On December 1, day breaks at 5h 40m and sunrise is at 7h 40m am. The Sun sets at 3h 55m and twilight ends at 5h 55m pm, the length of actual daylight being about 8 hours 10 minutes. During the month, the mornings decrease by 23 minutes and the afternoons increase by 5 minutes.

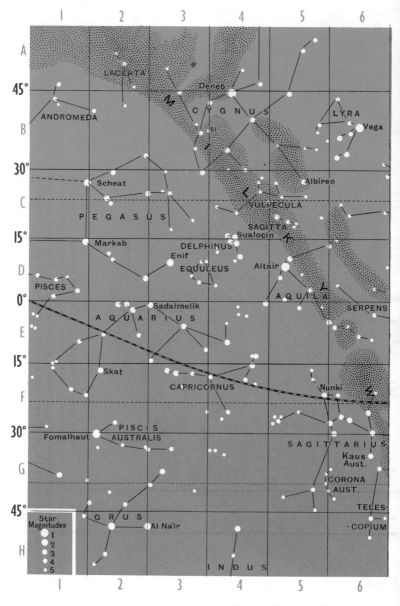

.... Southern limit of the Heavens visible from around latitude 51.5°N

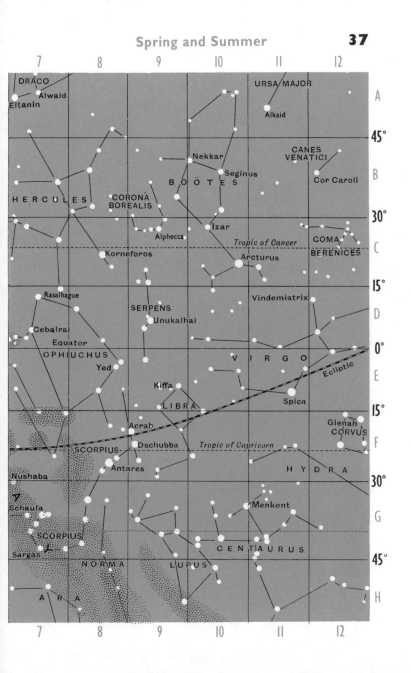

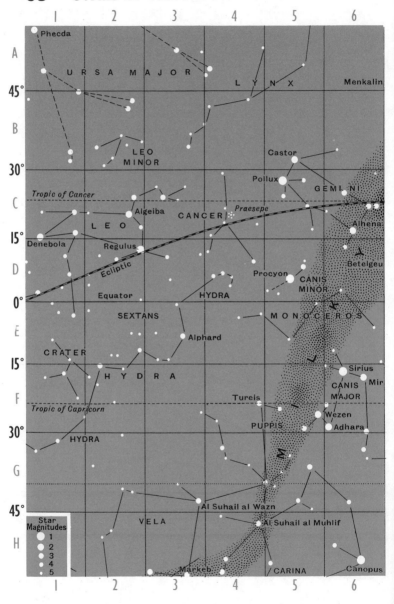

Phecda

A

URSA MAJOR

45°

LYNX

Menkalin

B

LEO
MINOR

Castor

30°

Pollux

GEMINI

Tropic of Cancer

C

Algeiba

CANCER

Praesepe

Alhena

LEO

15°

Denebola

Regulus

Betelgeu

Ecliptic

D

Procyon

CANIS
MINOR

0°

Equator

HYDRA

SEXTANS

MONOCEROS

E

Alphard

15°

CRATER

Sirius

HYDRA

CANIS
MAJOR

Mir

F

Tureis

Tropic of Capricorn

PUPPIS

Wezen

30°

HYDRA

Adhara

G

45°

Al Suhail al Wazn

VELA

Al Suhail al Muhlif

H

Star
Magnitudes
1
2
3
4
5

Markeb

CARINA

Canopus

.... *Southern limit of the Heavens visible from around latitude 51.5°N*

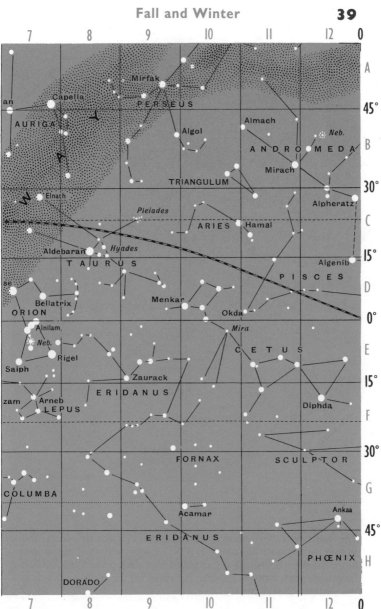

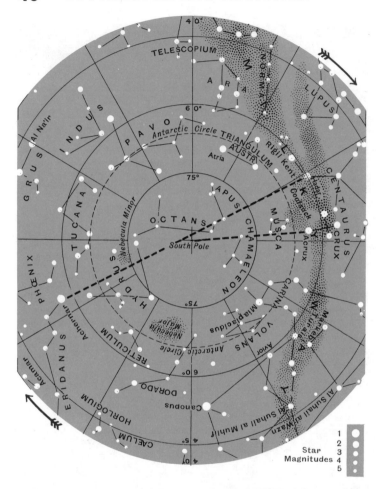

The stars shown above comprise all constellations which are *never visible* to the observer at the latitude of 51.5°N. The stars indicating the position of the Southern Celestial Pole are linked by a black pecked line; the Pole is approximately halfway between Achernar and the north–south axis of the Southern Cross (Crux).

Notes on the constellations and principal stars

The reference gives the page number of the monthly Chart on which the constellation first appears, if any, and its direction, and of the detailed charts (pages 9, 36–40), followed by the section. Numerals in brackets after the names of the stars indicate their magnitudes.

ANDROMEDA *12* **WNW;** *38–39* **B11**
Principal stars: Alpheratz (2.1), Mirach (2), Almach (2).
Andromeda is immediately below Cassiopeia, in line with it and the Pole Star. It is most conspicuous in the fall and early winter, but is visible nearly all year round. Alpheratz forms the upper left-hand corner of the Square of Pegasus. A remarkable nebula, visible to the naked eye in clear weather, is included in this star group.

APUS – The Bird of Paradise *40*

AQUARIUS – The Waterman or Water Bearer *23* **ESE;** *36–37* **E2**
Principal stars: Sadalmelik (3), Skat (3).
An extensive constellation below Pegasus; contains no notable stars.

AQUILA – The Eagle *21* **E;** *36–37* **D5**
Principal star: Altair (0.9).
Altair has a clear, bluish-white radiance, seen to best advantage in the evenings of early fall, when the star is midway between the zenith and the horizon. It lies between two fairly bright stars, and forms with them a straight line, which, if produced across the Milky Way, leads to Vega. Altair, Vega (in Lyra) and Deneb (in Cygnus) form a great triangle worth memorizing as a guide to other groups. Altair is the apex of the triangle.

ARA – The Altar *36–37* **H7;** *40*

ARIES – The Ram *13* **W;** *38–39* **C10**
Principal star: Hamal (2.2).
Two insignificant groups of these stars may be seen in winter below Andromeda, on a line from Cassiopeia through Almach. The lower is Aries and the upper TRIANGULUM, easily recognizable by its shape.

AURIGA – The Charioteer *13* **SW;** *9;* *38–39* **B7**
Principal stars: Capella (The Little Goat) (0.2), Menkalinan (2).
Auriga lies midway between Orion and the Pole Star, and is therefore most conspicuous in winter, when it crosses the meridian near the zenith in the wake of Cassiopeia. Capella is in line with Megrez and Dubhe (Big Dipper); it has an intense yellowish brilliance, and never sets.

BOÖTES – The Herdsman or Ox-driver *12* **NE;** *36–37* **B10**
Principal stars: Arcturus (0.2), Nekkar (3), Seginus (3).
The curve of the "handle" of the Big Dipper is continued by the stars of

the right side of Boötes, passes through Arcturus and terminates in Spica (Virgo). Boötes is characteristic of the short summer nights, when Arcturus, a bright reddish-yellow star, is the most striking object in the southern aspect of the heavens, and the third brightest star in the northern hemisphere.

CAELUM – The Sculptor's Chisel   *40*

CANCER – The Crab   *38–39* **C4**

A group of inconspicuous stars between Regulus (Leo) and Pollux (Gemini). It contains the famous cluster known as Praesepe, or the Beehive.

CANES VENATICI – The Greyhounds or Hunting Dogs   *36–37* **B12**

Cor Caroli (2) is the only noticeable star in this constellation, which is situated in a part of the sky, otherwise destitute of bright stars, between Virgo and the "handle" of the Big Dipper. There are known to be at least 30,000 stars in this cluster.

CANIS MAJOR – The Great Dog   *13* **SSE;** *38–39* **F6**

Principal stars: Sirius (−1.6), Mirzam (2), Adhara (1.6), Wezen (2).

Canis Major rises in the southeast, sets in the southwest, and reaches its greatest altitude in midwinter, when it is below and to the left of Orion. Sirius, "the blazing Dog Star," shines with an intense white light near the southern horizon during the winter months. It is the brightest star in the heavens, and one of the nearest to the Earth.

CANIS MINOR – The Little Dog   *13* **SE;** *38–39* **D5**

Canis Minor lies to the left of Orion and in line with Bellatrix and Betelgeuse; being higher in the sky, it is visible for a longer period than Canis Major. Procyon (0.5), a pale yellow star, forms with Sirius and Betelgeuse a great equilateral triangle which is one of the most striking features of the southern sky in winter.

CAPRICORNUS – The Horned Goat or Sea Goat   *25* **SE;** *36–37* **F3**

Its position near the horizon is indicated by the line of the three bright stars in Aquila. The most westerly star of the group can be resolved with the unaided eye into a pair, both of which have, in fact, fainter companions.

CARINA – The Keel   *40*

Principal stars: Canopus (−0.9), Miaplacidus (1.8), Tureis (2), Avior (2).

This large constellation, including Canopus, the second brightest star in the sky, is too far south to be seen from latitude 50°N.

CASSIOPEIA   *12* **NW;** *9*

Principal stars: Schedar (2.1 to 2.6), Caph (2).

This group resembles a distorted letter W drawn slantwise across the Milky Way, on the opposite side of the Pole Star from the Big Dipper. Cassiopeia is overhead in the fall and early winter, and being always visible and easily recognized is a useful guide to the location of less familiar star groups.

CENTAURUS – The Centaur   *36–37* G10; *40*

Rigil Kent (0.1), a navigation star, is the nearest star but one to the Earth, only four light-years away. It is a double star, though the two cannot be distinguished by the naked eye. Hadar (0.9) and Menkent (3) are also navigation stars.

CEPHEUS   *12* NNW; *9*

Alderaimin (2) lies between Deneb (Cygnus) and the Pole Star.

CETUS – The Whale or Sea Monster   *29* SE; *38–39* E11

Principal stars: Diphda (2.2), Menkar (2), Mira (5).

Rises due east and crosses the southern sky in advance of Orion, in the fall. The only interesting feature is Mira, a star which varies from third to ninth magnitude in the comparatively long period of 11 months, during half of which time it is invisible to the naked eye.

CHAMAELEON – Chameleon   *40*

COLUMBA – Noah's Dove (*see Lepus*)   *13* S; *38–39* G7

COMA BERENICES – Berenice's Hair (*see Virgo*)   *36–37* C12

CORONA AUSTRALIS – The Southern Crown   *36–37* G6

Another name for this constellation is Corolla.

CORONA BOREALIS – The Northern Crown   *14* NE; *36–37* B9

Alphecca (2.3) is the brightest of this semicircle of small stars, between Boötes and Hercules.

CORVUS – The Crow or Raven   *17* SSE; *36–37* F12

Gienah (2) and three other bright stars form a lozenge-shaped figure below Virgo. They must be sought low in the south in spring. Six stars grouped to the right of Corvus represent CRATER.

CRATER – The Cup (*see Corvus*)   *38–39* E1

CRUX – The Southern Cross   *40*

Principal stars: Acrux (1), Gacrux (2) – *see note on page 40*.

Close to the Cross there is a gap in the Milky Way which appears devoid of stars; this is the Coal Sack, a dark nebula which blocks the light of stars beyond.

CYGNUS – The Swan or Northern Cross   *12* NNW; *36–37* B4

Principal stars: Deneb (1.3), Albireo (3).

Cygnus is a striking feature of the heavens during late summer and fall when it passes directly overhead, and is recognizable as a great cross lying on the Milky Way, near Vega. Deneb is less brilliant than most first-magnitude stars. The small star marked 61 on page 36 is one of the nearest stars to the Earth, visible from northern latitudes. It is 64,800,000 million miles [104,300,000 million km] away, the first stellar distance measured (1838).

DELPHINUS – The Dolphin   *20* **ENE;** *36–37* **D4**

The Dolphin is a compact group of four stars forming a small lozenge, with a fifth below it, lying close to the Milky Way, and almost in line with the two lower stars of the Square of Pegasus. The figure cannot be mistaken when the sky is clear. EQUULEUS is a group of five fainter stars nearer to Pegasus.

DORADO – The Swordfish   *38–39* **H8;** *40*

DRACO – The Dragon   *12* **N;** *9*; *36–37* **A7**

Principal stars: Eltanin (Etamin) (2), Alwaid (3), Thuban (4).

Draco is a stream of fairly bright stars extending from close to Hercules up to the border of Cepheus, and thence curving back round Ursa Minor to a point between the Pole Star and the Pointers. Its brightest stars are grouped about a line drawn from Megrez (Big Dipper) to Vega (Lyra).

EQUULEUS – The Horse's Head (*see Delphinus*)   *36–37* **D3**

ERIDANUS – The River Eridanus or Po   *31* **ESE;** *38–39* **F8**

Principal stars: Achernar (0.6), Cursa (3), Zaurack (3), Acamar (3).

Eridanus is a long winding stream of stars which begins close to the right of Rigel (Orion), flows west toward Cetus, and curving back thence is finally lost to sight below the southern horizon.

FORNAX – The Furnace   *38–39* **G10**

GEMINI – The Twins   *13* **SE;** *38–39* **C5**

Principal stars: Pollux (1.2), Castor (1.6), Alhena (1.9).

Castor, a white star, and Pollux, with a golden tinge, represent the heads of the Twins at the eastern end of the constellation, which rises higher in the sky than Orion, in line with Betelgeuse and Rigel. In a telescope Castor resolves itself into three pairs of twin stars. Between the Twins and Polaris, the sky is empty of bright stars.

GRUS – The Crane   *36–37* **H2;** *40*

Principal star: Al Na'ir (2).

HERCULES   *14* **NE;** *36–37* **B7**

A summer constellation; the four brightest stars form an irregular quadrilateral to the left of Corona, on a line between Vega and Arcturus, and include an unusually red star.

HORLOGIUM – The Clock   *40*

HYDRA – The Water Serpent   *13* **ESE;** *38–39* **F2**

Alphard (The Solitary) (2.2) was so named on account of the dearth of bright stars in its vicinity. Hydra extends from the compact group of four stars just below Cancer to a star beneath Spica, and is the largest constellation in the heavens.

HYDRUS – The Water Snake   *40*

INDUS – The Indian   *36–37* **H4;** *40*

LACERTA – The Lizard   *36–37* **A2**

LEO – The Lion   15 ESE; 38–39 C2
Principal stars: Regulus or Cor Leonis (1.3), Denebola (2.2), Algeiba (2).
Leo is typical of the night sky of spring, when Orion is setting in the west. Megrez and Phecda (Big Dipper) point to Regulus, the "handle" of the Sickle. The stars of the latter group closely resemble the form assigned to them and are easily recognized. The remaining part of Leo, shaped like a triangle, lies some distance to the left. A feature of Leo is the meteoric shower occurring every November, when the Leonids appear to issue from a point in the sky close to the Sickle.

LEO MINOR – The Little Lion   38–39 B2

LEPUS – The Hare   13 S; 38–39 F7
The small star group immediately below the "feet" of Orion, south of Lepus, is COLUMBA, which is only seen for a brief period in winter.

LIBRA – The Balance or Scales   19 SE; 36–37 E9
Principal star: Zuben'ubi (3).
Visible in summer, low in the sky beneath Arcturus.

LUPUS – The Wolf   36–37 H9

LYNX – The Lynx   9; 38–39 A4

LYRA – The Lyre   16 NE; 9; 36–37 B6
Principal stars: Vega (0.1), Sulaphat (3).
Vega is a brilliant bluish-white star prominent near the zenith on summer nights, when Capella is low in the north. The Pole Star is about halfway between Vega and Capella, the three making a great horizontal line across the heavens in the fall and spring. Vega dips below the northern horizon of an observer in the northern USA, but elsewhere it is a true circumpolar star and never sets. With Altair and Deneb it forms a conspicuous V, with Vega at the top of the right arm of the V.

MONOCEROS – The Unicorn   38–39 E5
A minor constellation extending across the Milky Way east of Orion.

MUSCA – The Fly   40

NORMA – The Square   36–37 H8; 40

OCTANS – The Octant   40
This constellation includes the South Celestial Pole.

OPHIUCHUS – The Serpent Bearer   19 ESE; 36–37 E7
Principal stars: Rasalhague (2.1), Cebalrai (2), Yed (3).
Ophiuchus is the widespread star group immediately below Hercules. Barnard's "Proper Motion Star" in this constellation is the nearest known star to the Earth. SERPENS can be traced from below Corona across the Serpent Bearer and into the Milky Way beside Aquila.

ORION – The Hunter   13 S; 38–39 D7
Principal stars: Betelgeuse (0.1 to 1.2), Rigel (0.3), Bellatrix (1.7), Alnitak (2), Saiph (2).

Orion, the most brilliant star group of all, and the chief guide to the stars of the southern aspect of the heavens, rises and sets due east and west. The straight line of the three bright stars in Orion's "belt," if produced to the right (northwest), leads to Aldebaran (Taurus); if extended an equal distance to the left (northeast), it reaches Sirius (Canis Major). Betelgeuse varies considerably in brilliance and has a reddish tinge, visibly contrasting with the hard white radiance of Rigel. Keen-sighted observers may detect the Great Nebula in Orion as a haze of light around the middle star of the "sword" which hangs from the Hunter's belt.

PAVO – The Peacock   *40*
Principal star: Peacock (2).

PEGASUS – The Winged Horse   *22* ENE; *36–37* C2
Principal stars: Algenib (2), Markab (2), Scheat (2), Enif (2).

The Great Square of Pegasus, a quadrilateral of four bright stars (Algenib, Markab and Scheat, with Alpheratz of Andromeda), is a striking feature of the skies of late summer, fall and early winter, and is a most useful guide to less conspicuous constellations at these seasons. When ascending the eastern sky, and again when descending toward the northwest, the "Square" becomes a broad diamond-shaped figure, but it is unmistakable at all times. Pegasus crosses the meridian in the fall somewhat higher in the sky than Orion at its culmination in midwinter. A line drawn from Algenib to Polaris (through Alpheratz in Andromeda and Caph in Cassiopeia) would roughly indicate the zero meridian from which right ascension is calculated.

PERSEUS   *12* WNW; *9*; *38–39* A9
Mirfak (1.9) is the brightest star, and Algol (2) is a well-known variable star which changes from second to fourth magnitude in less than three days. Perseus is a festoon of stars on the Milky Way, between Capella and Cassiopeia.

PHOENIX   *38–39* H12; *40*
Principal star: Ankaa (Nair al Zaurak) (2).

PISCES – The Fishes   *38–39* D11
Principal star: Okda (4).

A line of stars extending from Aquarius below the Square of Pegasus to Aries.

PISCIS AUSTRALIS – The Southern Fish   *29* S; *36–37* F2
Fomalhaut (1.3), the one bright star, is visible for a short time in the fall when the horizon is clear. It should then be looked for in the south, below Aquarius, and in line with Scheat and Markab, the right-hand side of the Square of Pegasus.

PUPPIS – The Poop   *38–39* F4

RETICULUM – The Net   *40*
SAGITTA – The Arrow   *36–37* **C4**
Consists of four inconspicuous stars on the Milky Way above Aquila. Higher still, between Sagitta and Cygnus, a scattered group of small stars marks the position of VULPECULA.
SAGITTARIUS – The Archer   *25* **S;** *36–37* **G5**
Principal stars: Kaus Australis (1.9), Nunki (2), Nushaba (3).
Visible in summer near the southern horizon, below Aquila. Contains rich star clouds, and includes the center of our Galaxy.
SCORPIUS – The Scorpion   *21* **SSE;** *36–37* **F8**
Principal stars: Antares (1.2), Dschubba (2), Acrab (2), Shaula (2).
A summer constellation, seen as a festoon of bright stars just above the horizon, below Ophiuchus and the Serpent. Antares, the brightest star, has a reddish light, and has one of the largest known star diameters. During the short, light nights of June, it is due south about midnight.
SCULPTOR – Sculptor's Tools   *38–39* **G12**
SERPENS – The Serpent (*see Ophiuchus*)   *19* **ESE;** *36–37* **D9**
Principal star: Unuk (2).
SEXTANS – The Sextant   *38–39* **E2**
TAURUS – The Bull   *13* **SW;** *38–39* **D8**
Principal stars: Aldebaran (1.1), Elnath (1.8), Alcyone (2).
Elnath is midway between Capella (Auriga) and Betelgeuse (Orion). Aldebaran, a brilliant red star, lies considerably to the right of this line in the direction marked by the "belt" of Orion. It forms with four smaller stars the cluster known as Hyades. Higher on the right are the Pleiades. This well-known cluster has six stars easily visible to the naked eye, the brightest being Alcyone. The Pleiades first appear as a compact scintillating group in the northeast on the dark nights following the autumnal equinox, while the Hyades are yet below the horizon.
TELESCOPIUM – The Telescope   *36–37* **G6; 40**
TRIANGULUM – The Triangle (*see Aries*)   *12* **W;** *38–39* **B10**
Principal star: Atria (2).
TRIANGULUM AUSTRALE – The Southern Triangle   *40*
TUCANA – The Toucan   *40*
URSA MAJOR – The Great Bear   *12* **NE;** *9*
Principal stars: Dubhe (1.9), Merak (2), Phecda (2), Megrez (3), Alioth (1.7), Mizar (2), Alkaid (Benetnasch) (1.9).
These seven stars form the Big Dipper (also known as the Plough and Charles's Wain), which is the most conspicuous and familiar star group of the northern hemisphere. A line drawn through Merak and Dubhe, the "Pointers," passes close to the Pole Star. Alkaid, the tail star of the Big Dipper handle, is a useful pointer to Arcturus to the south. The Big Dipper

never sets, and is always high enough in the sky to be easily seen; it is therefore the most useful guide to the location of the North Celestial Pole. In winter, the Big Dipper is in the northeast, with the Pointers uppermost. During March and April it passes close to the zenith, and descending in the northwest as summer advances, it reaches a position between the Pole Star and northern horizon in the evenings of October and November. An interesting feature of the group is the small star Alcor, a companion of Mizar, visible to keen sight on a clear night.

URSA MINOR – The Little Bear *12* **N;** *9*
Principal stars: Polaris (Pole Star) (2.1), Kochab (2).
The Pole Star is so close to the North Celestial Pole that its own revolution is neglible and it is accepted as indicating at all times the true north. It is thus the most important guide star of the night sky. The Little Bear is otherwise an inconspicuous group, and can be traced as a faint string of stars between Polaris and Kochab, with a small rectangle at the latter end. The line Polaris–Kochab points to Boötes.

VELA – The Sails *38–39* **H3**
Principal stars: Al Suhail al Muhlif (2), Al Suhail al Wazn (2).

VIRGO – The Virgin *15* **ESE;** *36–37* **E11**
Principal stars: Spica (1.2), Vindemiatrix (2).
Virgo is characteristic of the nights of early spring, and crosses the meridian in March. Spica lies at the end of the curve of the "handle" of the Big Dipper, continued through Arcturus. Spica and Arcturus with Denebola form a nearly equilateral triangle. A line through Spica and Vindemiatrix to the Big Dipper crosses a portion of the sky deficient in bright stars. Cor Caroli (Canes Venatici) is the only prominent star near this line, and between it and Vindemiatrix is a group of very faint stars known as COMA BERENICES.

VOLANS – The Flying Fish *40*
VULPECULA – The Fox (*see Sagitta*) *36–37* **C4**

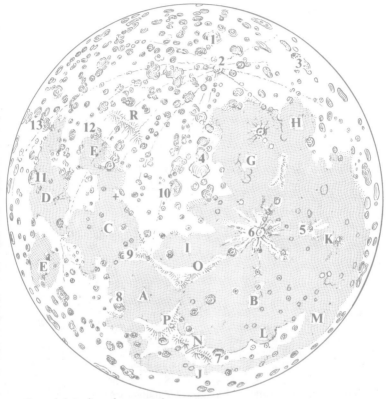

Seas (plains) and mountains

A	Mare Serenitatis	G	Mare Nubium	M	Sinus Roris
B	Mare Imbrium	H	Mare Humorum	N	The Alps
C	Mare Tranquilitatis	I	Mare Vaporum	O	The Apennine Mountains
D	Mare Fecunditatis	J	Mare Frigoris	P	The Caucasus Mountains
E	Mare Crisium	K	Oceanus Procellarum	R	The Altai Mountains
F	Mare Nectaris	L	Sinus Iridum		

Principal craters

1	Clavius	6	Copernicus	11	Langrenus
2	Tycho	7	Plato	12	Fracastorius
3	Schickard	8	Posidonius	13	Petavius
4	Alphonsus	9	Plinius		
5	Kepler	10	Hipparchus		

+   Apollo 11 landing site

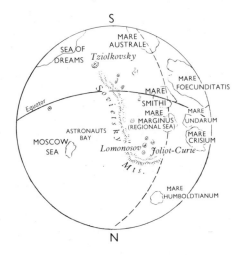

Planetary distances

Name	Distance from the Sun	
	(miles)	**(kilometers)**
Mercury	36 million miles	58 million km
Venus	67 million miles	108 million km
Earth	93 million miles	150 million km
Mars	142 million miles	228 million km
Jupiter	484 million miles	778 million km
Saturn	887 million miles	1430 million km
Uranus	1780 million miles	2870 million km
Neptune	2800 million miles	4500 million km
Pluto	3700 million miles	5900 million km

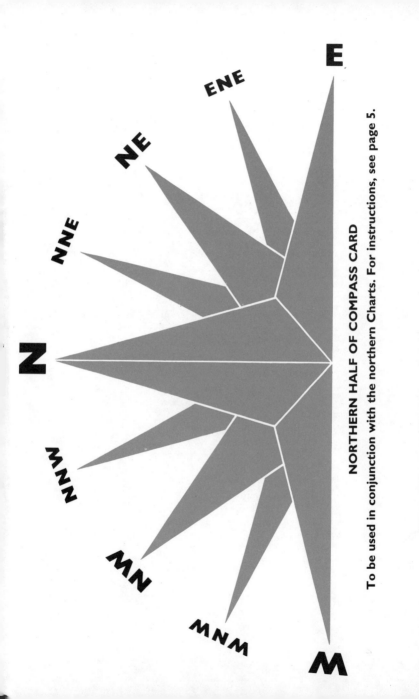

NORTHERN HALF OF COMPASS CARD

To be used in conjunction with the northern Charts. For instructions, see page 5.